~A BINGO BOOK~

Oceans and Marine Life Bingo Book

COMPLETE BINGO GAME IN A BOOK

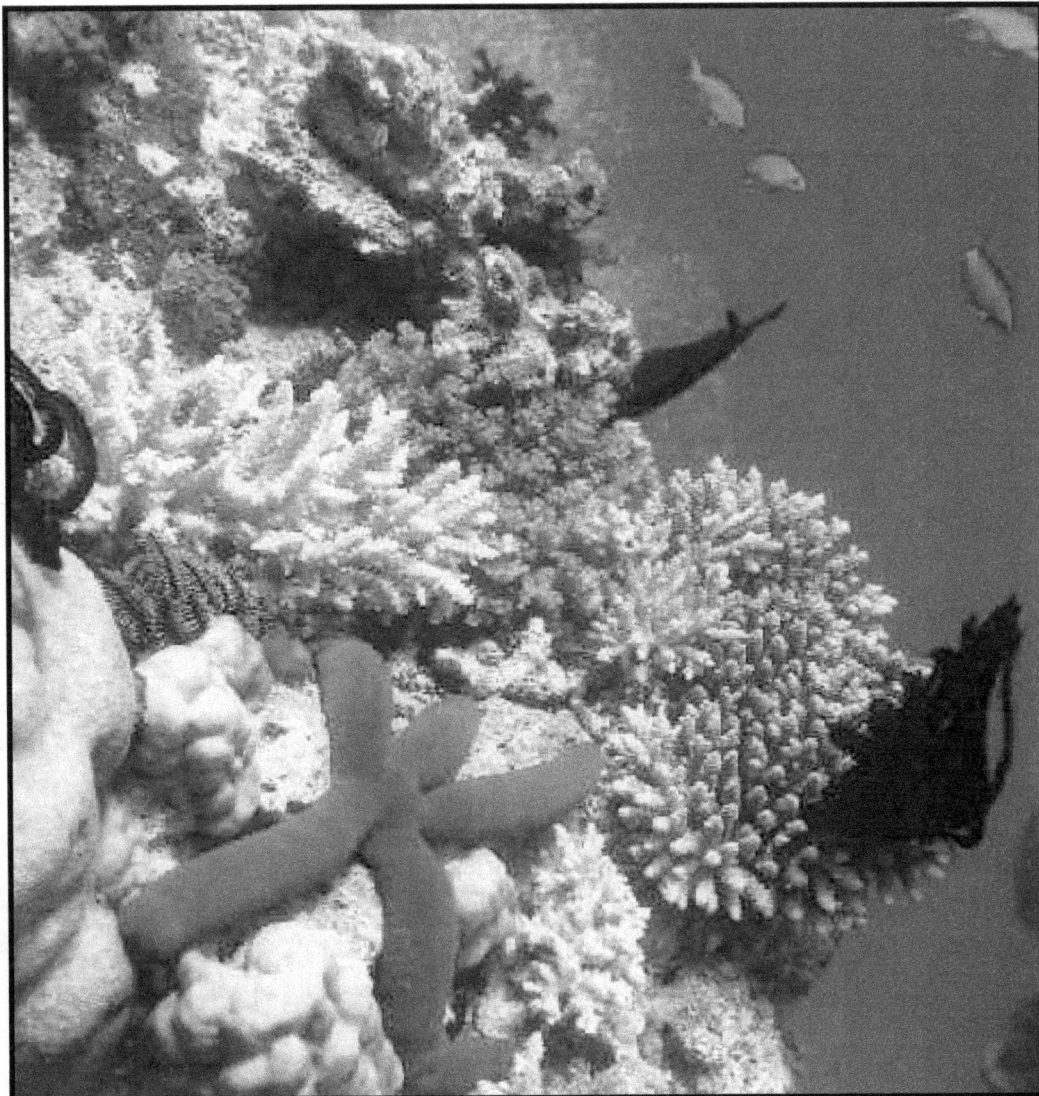

Photo is Copyright (c) 2004 Richard Ling

Written By Rebecca Stark

ISBN 978-0-87386-438-1

Educational Books 'n' Bingo

Printed in the U.S.A.

OCEANS & MARINE LIFE BINGO DIRECTIONS

INCLUDED:

List of Terms

Templates for Additional Terms and Clues

2 Clues per Term

30 Unique Bingo Cards

Markers

1. **Either cut apart the book or make copies of ALL the sheets. You might want to make an extra copy of the clue sheets to use for introduction and review. Keep the sheets in an envelope for easy reuse.**

2. Cut apart the call cards with terms and clues.

3. Pass out one bingo card per student. There are enough for a class of 30.

4. Pass out markers. You may cut apart the markers included in this book or use any other small items of your choice.

5. Decide whether or not you will require the entire card to be filled. Requiring the entire card to be filled provides a better review. However, if you have a short time to fill, you may prefer to have them do the just the border or some other format. Tell the class before you begin what is required.

6. There are 50 terms. Read the list before you begin. If there are any terms that have not been covered in class, you may want to read to the students the term and clues before you begin.

7. There is a blank space in the middle of each card. You can instruct the students to use it as a free space or you can write in answers to cover terms not included. Of course, in this case you would create your own clues. (Templates provided.)

8. Shuffle the cards and place them in a pile. Two or three clues are provided for each term. If you plan to play the game with the same group more than once, you might want to choose a different clue for each game. If not, you may choose to use more than one clue.

9. Be sure to keep the cards you have used for the present game in a separate pile. When a student calls, "Bingo," he or she will have to verify that the correct answers are on his or her card AND that the markers were placed in response to the proper questions. Pull out the cards that are on the student's card keeping them in the order they were used in the game. Read each clue as it was given and ask the student to identify the correct answer from his or her card.

10. If the student has the correct answers on the card AND has shown that they were marked in response to the *correct questions,* then that student is the winner and the game is over. If the student does not have the correct answers on the card OR he or she marked the answers in response to *the wrong questions,* then the game continues until there is a proper winner.

11. If you want to play again, reshuffle the cards and begin again.

Have fun!

TERMS

algae (singular is alga)

aphotic

Arctic Ocean

arthropod(s)

Atlantic Ocean

benthic

bioluminescence

bony fish

buoyancy

cartilaginous fish

cetaceans

coral

current(s)

epipelagic zone

estuary

food chain

freshwater

Indian Ocean

jellyfish

krill

intertidal zone

marine

mid-ocean ridge

mollusks

nekton

ocean

oceanography

Pacific Ocean

photic

photosynthesis

pinnipeds

plankton

predator(s)

prey

reef(s)

Ring of Fire

salinity

Sargasso Sea

scales

school(s)

scuba

sharks

Southern Ocean

spawn

sponges

tectonic plates

tidal pool(s)

tide(s)

tsunami(s)

wave

Additional Terms

Choose as many terms as you would like and write them in the squares.
Repeat each as desired. Cut out the squares and randomly
distribute them to the class.
Instruct the students to place the square on the center space of their card.

Oceans & Marine Life Bingo

© Barbara M. Peller

Clues for Additional Terms

Write two or three clues for each new term.

_____ 1. _____ 2. _____ 3. _____	_____ 1. _____ 2. _____ 3. _____
_____ 1. _____ 2. _____ 3. _____	_____ 1. _____ 2. _____ 3. _____
_____ 1. _____ 2. _____ 3. _____	_____ 1. _____ 2. _____ 3. _____

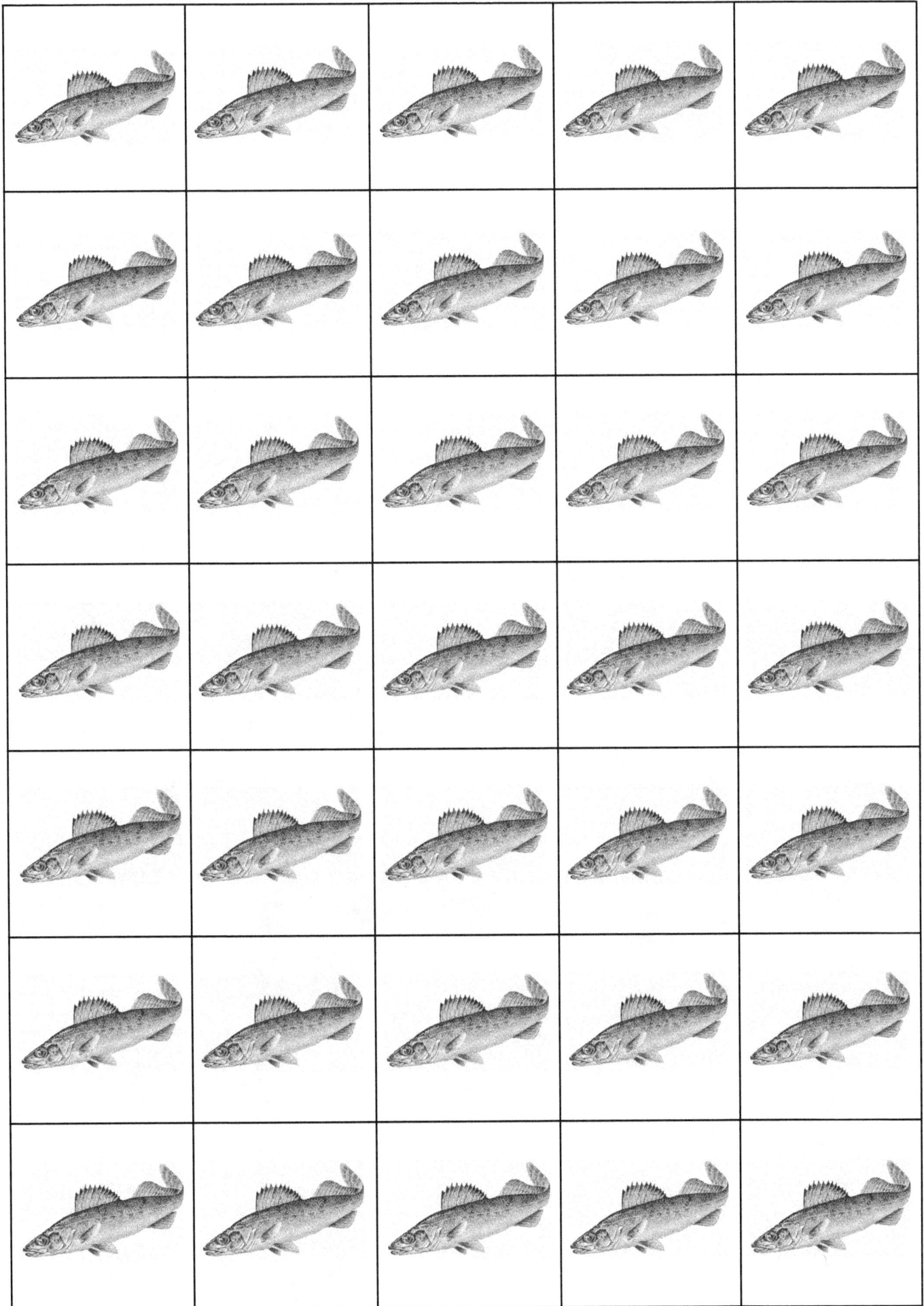

algae

1. ___ are plant-like, mainly aquatic, organisms. Unlike plants, they lack true stems, leaves and roots. Kelp and other seaweeds are types of ___.

2. ___ are autotrophic, which means they get their energy through photosynthesis.

aphotic

1. The ___ zone is the part of a lake or ocean where there is little or no sunlight. It is the opposite of the photic, or euphotic, zone.

2. No photosynthesis occurs in the ___ zone of the ocean because no sunlight reaches that region.

Arctic Ocean

1. The ___ is the smallest and most shallow ocean. It is partly covered with sea ice throughout the year.

2. The North Pole is in the middle of the ___.

arthropod(s)

1. ___ are invertebrates with jointed limbs, a segmented body, and an exoskeleton made of chitin.

2. More than 80% of all the animals on Earth are ___, including lobsters, crabs, spiders, shrimps, and barnacles.

Atlantic Ocean

1. The ___ is the second largest ocean, but it is relatively shallow.

2. The ___ is located between the continents of North America and Europe in the north and South America and Africa in the south.

benthic

1. ___ refers to aquatic organisms that live in, on, or near the bottom of a body of water.

2. Creeping animals, such as snails; burrowing animals, such as many clams; and animals unable to move about, such as corals and sponges, are ___ animals.

bioluminescence

1. Light produced by a chemical reaction within an organism is called ___.

2. Through a process known as ___, some deep-sea animals use chemicals within their bodies to produce light.

bony fish

1. The name of their class, Osteichthyes, comes from two Greek words: *osteon,* meaning "bone," and *icthys,* meaning "fish."

2. There are more than 29,000 species of these vertebrates. Salmon and tuna are examples.

buoyancy

1. The ability of something to float is called ___.

2. The upward pressure exerted upon a floating body by a liquid is called ___.

cartilaginous fish

1. The internal skeleton of these aquatic vertebrates is made entirely of cartilage.

2. Vertebrates such as sharks, skates, and rays are ___.

cetaceans	coral
1. These large, aquatic, carnivorous mammals have fin-like forelimbs and no hind limbs. 2. These large, aquatic, carnivorous mammals include whales, dolphins, and porpoises.	1. A ___ reef is formed from tiny animals that live in colonies; these organisms are called ___ polyps. 2. ___ reefs develop in shallow, warm water and are found mostly in the tropics. These ocean habitats are rich in life.
current(s)	**epipelagic zone**
1. The steady flow of the ocean's surface water in a prevailing direction is called an ocean ___. 2. The Gulf Stream is a large warm ocean ___ that flows northward from the Gulf of Mexico.	1. The top layer of the ocean is called the epipelagic, or euphotic, zone. 2. Most life is in the ___ because it is the sunlit zone and photosynthesis can take place.
estuary	**food chain**
1. An ___ forms where freshwater from rivers and streams flows into the ocean. 2. In an ___ freshwater mixes with seawater.	1. The term ___ is used to describe the feeding relationships in a community of organisms. 2. Phytoplankton, a primary producer, is at the first level of the marine ___.
freshwater	**Indian Ocean**
1. ___ has a low salt concentration. 2. Plants and animals that inhabit ponds, lakes, streams, rivers and other ___ areas would not be able to survive in waters with a high salt concentration.	1. The ___ is located between Africa, the Southern Ocean, Asia and Australia. 2. Madagascar and Sri Lanka are island countries in the ___.
jellyfish	**krill**
1. These carnivorous invertebrates have stinging cells and a saclike body cavity. 2. Tentacles with stinging cells dangle from their bell-shaped bodies.	1. ___ are small, shrimp-like crustaceans. They are a type of zooplankton. 2. These shrimp-like marine invertebrates feed mostly on phytoplankton.

intertidal zone 1. The ___ is the region where the land and sea meet. It is exposed to the air at low tide and covered with water at high tide. 2. The ___ is sometimes called the littoral zone, although they are not exactly the same.	**marine** 1. The term ___ means "of or relating to the sea." 2. The study of organisms that live in the ocean is called ___ biology.
mid-ocean ridge 1. A ___ is an underwater volcanic mountain range on an ocean floor formed by plate tectonics. 2. The ___ has a valley known as a rift running down its center.	**mollusks** 1. ___ are marine animals with a soft, unsegmented body enclosed in a shell. 2. Clams, snails, scallops and oysters are types of ___.
nekton 1. ___ comprises animals that move freely. 2. Whales, fish and sea turtles are al considered ___ because they move freely.	**ocean** 1. The ___ is divided into five major divisions: the Arctic, the Atlantic, the Indian, the Pacific, and the Southern. 2. The body of salt water that covers more than 70 percent of Earth's surface is called the ___.
oceanography 1. The scientific study of the ocean and its phenomena is called ___. 2. A scientist who specializes in this branch of science is called an oceanographer.	**Pacific Ocean** 1. The ___ is the largest and the deepest ocean. 2. The longest coral reef in the world, the Great Barrier Reef, is located in the ___.
photic 1. The ___ zone is the depth in the ocean or a lake where photosynthesis can occur. 2. The ___, or euphotic, zone receives enough sunlight for photosynthesis to occur. Oceans and Marine Life Bingo	**photosynthesis** 1. ___ is the process by which green plants synthesize carbohydrates from carbon dioxide and water using light energy 2. Most phytoplankton is found near the surface of the water, where there is enough sunlight for ___ to take place. © Barbara M. Peller

pinnipeds 1. ___ are fin-footed, semi-aquatic marine mammals. 2. Seals, sea lions and walruses are ___.	**plankton** 1. ___ is the collective name for the microscopic plants and animals that drift in the water. 2. Algae are forms of phyto___. Jellyfish and krill are forms of zoo___.
predator(s) 1. An organism that survives by preying on other animals is a ___. 2. Large ___ are at the top of the marine food chain. Some of those ___ include seals, tuna and pelicans.	**prey** 1. Albacore tuna are top predators and often ___ on schools of smaller fish, such as sardines and anchovies. 2. An animal taken by a predator for food is called the ___.
reef(s) 1. An atoll is a ring of coral ___ that encircle a lagoon. 2. There are three main types of coral ___: fringing ___, barrier ___, and atolls.	**Ring of Fire** 1. Many earthquakes and volcanoes occur in the Pacific Basin, known as the ___. 2. The Pacific Ocean has a basin rim that is ringed with volcanoes. That is why the Pacific Basin is often called the ___.
salinity 1. ___ refers to the concentration of salt in the oceans, seas, lakes, and rivers. 2. The Great Salt Lake of Utah has an even higher ___ than sea water.	**Sargasso Sea** 1. The ___ is in the North Atlantic Ocean. It is so named because of the seaweed that floats on the surface. 2. The seaweed that floats on the surface of the ___ is of the genus Sargassum.
scales 1. These small, plate-like structures form the external covering of many fish. 2. A fish's ___ protect it from its environment. The ___ overlap so that the fish's skin is not exposed.	**school(s)** 1. A large number of the same kind of fish swimming together is called a ___. 2. The main reason fish travel in ___ is for protection.

© **Barbara M. Peller**

scuba 1. ___ is an acronym for **s**elf-**c**ontained **u**nderwater **b**reathing **a**pparatus. 2. This equipment is worn by divers for breathing underwater.	**sharks** 1. Along with skates and rays, ___ are cartilaginous fish. 2. The mako and great white are two types of ___.
Southern Ocean 1. Until 2000 the ___ was considered to be parts of the Atlantic, Pacific and Indian oceans. 2. The ___ surrounds Antarctica and is the only ocean to surround a continent.	**spawn** 1. Aquatic animals are said to ___ when they produce or deposit eggs. 2. The eggs of fish and other aquatic animals are referred to as ___.
sponges 1. ___ are primitive life forms. The phylum to which they belong, Porifera, means "pore-bearer." 2. ___ are porous. This allows the water to flow in and out of their bodies.	**tectonic plates** 1. Earth's lithosphere (crust and upper mantle) are broken up into ___. 2. The location where two ___ meet is called a plate boundary.
tidal pool(s) 1. A pool of water that remains on a shore or reef after the tide recedes is called a ___. 2. ___ form along the shore when high tide comes in and fills in the holes and cracks of the rocky area. With low tide, most of the water returns to the ocean.	**tide(s)** 1. The alternate rising and falling of the surface of the ocean are called ___. 2. There are two each day: high ___ and low ___.
tsunami 1. A ___ is a destructive sea wave caused by an earthquake or a volcanic eruption. 2. These destructive waves are often mistakingly called tidal waves, but they have nothing to do with tides. Oceans and Marine Life Bingo	**wave** 1. One of the series of ridges that move across the surface of the ocean is called a ___. 2. A tsunami is a huge, destructive one.

Oceans and Marine Life Bingo

Indian Ocean	Arctic Ocean	marine	tidal pool(s)	sharks
buoyancy	jellyfish	sponges	Pacific Ocean	ocean
scales	predator(s)		intertidal zone	photosynthesis
tectonic plates	aphotic	freshwater	wave	mid-ocean ridge
mollusks	tsunami	cetaceans	algae	krill

Oceans and Marine Life Bingo

tidal pool(s)	scuba	nekton	plankton	mollusks
mid-ocean ridge	estuary	bony fish	aphotic	Sargasso Sea
reef(s)	tsunami		coral	freshwater
Pacific Ocean	Ring of Fire	predator(s)	tide(s)	ocean
krill	sponges	cetaceans	buoyancy	algae

Oceans and Marine Life Bingo: Card No. 2

Oceans and Marine Life Bingo

tidal pool(s)	freshwater	Pacific Ocean	wave	scales
tsunami	Arctic Ocean	Atlantic Ocean	jellyfish	photic
aphotic	sponges		salinity	arthropod(s)
predator(s)	reef(s)	mollusks	estuary	nekton
algae	cetaceans	buoyancy	tide(s)	marine

Oceans and Marine Life Bingo

predator(s)	salinity	marine	cetaceans	mollusks
oceanography	estuary	jellyfish	plankton	scales
intertidal zone	bony fish		sharks	wave
freshwater	food chain	sponges	buoyancy	Atlantic Ocean
algae	krill	pinnipeds	epipelagic zone	photosynthesis

Oceans and Marine Life Bingo

krill	sharks	aphotic	bony fish	cetaceans
oceanography	freshwater	Atlantic Ocean	predator(s)	current(s)
scuba	photosynthesis		Arctic Ocean	marine
ocean	salinity	Indian Ocean	tide(s)	epipelagic zone
Pacific Ocean	buoyancy	prey	coral	intertidal zone

Oceans and Marine Life Bingo

arthropod(s)	salinity	nekton	scuba	photosynthesis
wave	aphotic	epipelagic zone	jellyfish	scales
plankton	Atlantic Ocean		bony fish	coral
buoyancy	mollusks	tide(s)	pinnipeds	intertidal zone
mid-ocean ridge	freshwater	Indian Ocean	prey	marine

Oceans and Marine Life Bingo

Indian Ocean	salinity	school(s)	current(s)	Pacific Ocean
mid-ocean ridge	marine	tsunami	Arctic Ocean	scales
nekton	wave		coral	benthic
predator(s)	estuary	oceanography	tidal pool(s)	reef(s)
cetaceans	buoyancy	tide(s)	pinnipeds	arthropod(s)

Oceans and Marine Life Bingo

intertidal zone	salinity	bioluminescence	wave	benthic
oceanography	scuba	plankton	marine	sharks
scales	Sargasso Sea		photosynthesis	bony fish
algae	predator(s)	tidal pool(s)	epipelagic zone	estuary
sponges	buoyancy	pinnipeds	aphotic	mid-ocean ridge

Oceans and Marine Life Bingo

coral	Pacific Ocean	tsunami	scales	photosynthesis
epipelagic zone	scuba	intertidal zone	aphotic	marine
photic	Indian Ocean		Arctic Ocean	bioluminescence
benthic	krill	mollusks	current(s)	school(s)
estuary	tide(s)	Atlantic Ocean	tidal pool(s)	sharks

Oceans and Marine Life Bingo

tectonic plates	tidal pool(s)	bony fish	plankton	prey
photosynthesis	benthic	jellyfish	Arctic Ocean	marine
salinity	Sargasso Sea		wave	reef(s)
mollusks	ocean	epipelagic zone	tide(s)	photic
cartilaginous fish	krill	nekton	mid-ocean ridge	intertidal zone

Oceans and Marine Life Bingo

arthropod(s)	Sargasso Sea	aphotic	epipelagic zone	mid-ocean ridge
bioluminescence	photic	current(s)	coral	jellyfish
oceanography	scuba		nekton	tsunami
cartilaginous fish	scales	tide(s)	buoyancy	tidal pool(s)
Atlantic Ocean	cetaceans	Indian Ocean	pinnipeds	Pacific Ocean

Oceans and Marine Life Bingo

Pacific Ocean	estuary	photic	wave	coral
tsunami	sponges	scuba	pinnipeds	oceanography
Indian Ocean	school(s)		photosynthesis	plankton
cetaceans	sharks	marine	tidal pool(s)	Arctic Ocean
Sargasso Sea	bioluminescence	salinity	Atlantic Ocean	benthic

Oceans and Marine Life Bingo

cartilaginous fish	sharks	arthropod(s)	photic	photosynthesis
scuba	bioluminescence	salinity	coral	reef(s)
wave	bony fish		tsunami	school(s)
intertidal zone	tide(s)	benthic	Sargasso Sea	tidal pool(s)
buoyancy	ocean	pinnipeds	Indian Ocean	current(s)

Oceans and Marine Life Bingo: Card No.13

© Barbara M. Peller

Oceans and Marine Life Bingo

cetaceans	scuba	aphotic	coral	cartilaginous fish
benthic	Indian Ocean	photic	Arctic Ocean	reef(s)
epipelagic zone	wave		nekton	bony fish
ocean	tide(s)	salinity	Atlantic Ocean	arthropod(s)
buoyancy	plankton	Sargasso Sea	mid-ocean ridge	intertidal zone

Oceans and Marine Life Bingo

current(s)	coral	aphotic	Pacific Ocean	marine
arthropod(s)	prey	jellyfish	scuba	epipelagic zone
photosynthesis	Indian Ocean		scales	wave
buoyancy	photic	bioluminescence	tide(s)	cartilaginous fish
mid-ocean ridge	estuary	pinnipeds	nekton	tsunami

Oceans and Marine Life Bingo

bony fish	spawn	bioluminescence	prey	Ring of Fire
plankton	Sargasso Sea	school(s)	oceanography	tectonic plates
cartilaginous fish	sharks		photosynthesis	tsunami
predator(s)	estuary	buoyancy	current(s)	tidal pool(s)
epipelagic zone	photic	pinnipeds	benthic	reef(s)

Oceans and Marine Life Bingo: Card No. 16

Oceans and Marine Life Bingo

cartilaginous fish	Southern Ocean	food chain	photic	buoyancy
current(s)	epipelagic zone	tide(s)	wave	school(s)
coral	tectonic plates		spawn	bioluminescence
krill	mid-ocean ridge	intertidal zone	aphotic	reef(s)
mollusks	Atlantic Ocean	Pacific Ocean	tidal pool(s)	sharks

Oceans and Marine Life Bingo

marine	salinity	benthic	epipelagic zone	plankton
krill	cartilaginous fish	aphotic	photosynthesis	Atlantic Ocean
coral	reef(s)		food chain	prey
Sargasso Sea	jellyfish	tide(s)	tectonic plates	nekton
spawn	photic	mollusks	Southern Ocean	arthropod(s)

Oceans and Marine Life Bingo

photosynthesis	arthropod(s)	photic	bioluminescence	Sargasso Sea
current(s)	cetaceans	prey	Pacific Ocean	tectonic plates
Southern Ocean	wave		Arctic Ocean	aphotic
nekton	spawn	mollusks	estuary	food chain
scales	Ring of Fire	mid-ocean ridge	intertidal zone	pinnipeds

Oceans and Marine Life Bingo

Sargasso Sea	Southern Ocean	tectonic plates	photic	Arctic Ocean
bony fish	tsunami	oceanography	mollusks	plankton
sharks	school(s)		predator(s)	food chain
krill	intertidal zone	algae	estuary	spawn
freshwater	sponges	Ring of Fire	tidal pool(s)	jellyfish

© Barbara M. Peller

Oceans and Marine Life Bingo

arthropod(s)	krill	oceanography	photic	ocean
sharks	food chain	benthic	bioluminescence	Indian Ocean
reef(s)	mid-ocean ridge		Southern Ocean	aphotic
mollusks	Pacific Ocean	spawn	current(s)	intertidal zone
predator(s)	Ring of Fire	pinnipeds	cartilaginous fish	estuary

Oceans and Marine Life Bingo

scales	nekton	food chain	scuba	cartilaginous fish
plankton	tectonic plates	marine	bioluminescence	Arctic Ocean
benthic	wave		Indian Ocean	school(s)
spawn	krill	estuary	jellyfish	cetaceans
Ring of Fire	Atlantic Ocean	Southern Ocean	reef(s)	oceanography

Oceans and Marine Life Bingo

bony fish	Southern Ocean	Pacific Ocean	scuba	pinnipeds
arthropod(s)	Sargasso Sea	mid-ocean ridge	current(s)	jellyfish
nekton	cartilaginous fish		algae	Indian Ocean
reef(s)	sponges	spawn	Atlantic Ocean	estuary
ocean	intertidal zone	Ring of Fire	mollusks	food chain

Oceans and Marine Life Bingo

bony fish	Sargasso Sea	cetaceans	Southern Ocean	bioluminescence
photosynthesis	pinnipeds	oceanography	plankton	Indian Ocean
school(s)	prey		cartilaginous fish	reef(s)
ocean	algae	spawn	Atlantic Ocean	sharks
freshwater	predator(s)	Ring of Fire	tectonic plates	sponges

Oceans and Marine Life Bingo: Card No. 24

Oceans and Marine Life Bingo

predator(s)	oceanography	Southern Ocean	aphotic	food chain
jellyfish	ocean	current(s)	bony fish	Arctic Ocean
sharks	bioluminescence		algae	spawn
prey	krill	sponges	Ring of Fire	tectonic plates
pinnipeds	cetaceans	benthic	epipelagic zone	freshwater

Oceans and Marine Life Bingo: Card No. 25

Oceans and Marine Life Bingo

food chain	Southern Ocean	algae	plankton	prey
nekton	wave	bioluminescence	Sargasso Sea	bony fish
ocean	mollusks		tectonic plates	predator(s)
cartilaginous fish	scuba	krill	Ring of Fire	spawn
school(s)	epipelagic zone	aphotic	sponges	freshwater

Oceans and Marine Life Bingo

algae	benthic	Southern Ocean	Sargasso Sea	tsunami
ocean	nekton	current(s)	spawn	Arctic Ocean
tide(s)	sponges		Ring of Fire	predator(s)
prey	arthropod(s)	freshwater	oceanography	jellyfish
cartilaginous fish	tectonic plates	food chain	scales	school(s)

Oceans and Marine Life Bingo

photosynthesis	salinity	tidal pool(s)	Southern Ocean	benthic
tsunami	food chain	algae	mollusks	tectonic plates
sponges	reef(s)		prey	plankton
school(s)	scales	mid-ocean ridge	Ring of Fire	spawn
scuba	coral	cartilaginous fish	freshwater	ocean

Oceans and Marine Life Bingo: Card No. 28

Oceans and Marine Life Bingo

food chain	salinity	prey	current(s)	coral
ocean	mollusks	oceanography	school(s)	scales
sharks	Southern Ocean		Arctic Ocean	algae
tsunami	krill	marine	Ring of Fire	spawn
bony fish	bioluminescence	freshwater	arthropod(s)	sponges

Oceans and Marine Life Bingo: Card No. 29

© Barbara M. Peller

Oceans and Marine Life Bingo

cetaceans	Southern Ocean	plankton	coral	spawn
jellyfish	prey	nekton	tectonic plates	Arctic Ocean
freshwater	sponges		school(s)	oceanography
ocean	arthropod(s)	food chain	Ring of Fire	algae
krill	Pacific Ocean	Atlantic Ocean	salinity	marine

Oceans and Marine Life Bingo: Card No. 30

www.ingramcontent.com/pod-product-compliance
Lightning Source LLC
Chambersburg PA
CBHW051419200326
41520CB00023B/7300